Imitating *Nature*

From Barbs on a Weed to VELCRO

Other books in this series include:

From Bat Sonar to Canes for the Blind
From Bug Legs to Walking Robots
From Spider Webs to Man-Made Silk
From Penguin Wings to Boat Flippers

Imitating *Nature*

From Barbs on a Weed to VELCRO

Toney Allman

KIDHAVEN PRESS
An imprint of Thomson Gale, a part of The Thomson Corporation

THOMSON
✦
GALE™

Detroit • New York • San Francisco • San Diego • New Haven, Conn. • Waterville, Maine • London • Munich

LIBRARY OF CONGRESS CATALOGING-IN-PUBLICATION DATA

Allman, Toney.
 From barbs on a weed to Velcro / by Toney Allman.
  p. cm. — (Imitating nature)
 Includes bibliographical references and index.
 ISBN 0-7377-3118-4 (hc. : alk. paper)
 1. Inventions—Juvenile literature. I. Title. II. Series.
 T48.A42 2005
 600—dc22
                        2004022980

Printed in The United States of America

Contents

CHAPTER 1

Pesky Weeds

One late summer day in Switzerland, George de Mestral came home from walking through the fields with his dog. At the end of their walk, he had an unpleasant job. Dozens of little **burs** had stuck to his dog's fur and to his own wool pants. Carefully, he plucked them off one by one. The chore was a nuisance, but de Mestral was fascinated by the little sticky burs and how they clung to things so tightly.

The Burdock Plant

The burs came from a plant that is a member of the sunflower family. No one is sure which member of the sunflower family it was. Some people think it was the cocklebur, but many people believe de Mestral's burs were from the burdock plant. Burdock grows wild in Europe, Asia, Canada, and the United States. It is called a weed because it has no value to people and often grows where it is not wanted.

The burdock bur (shown magnified, inset) and the cocklebur (main) come from weeds that belong to the sunflower family.

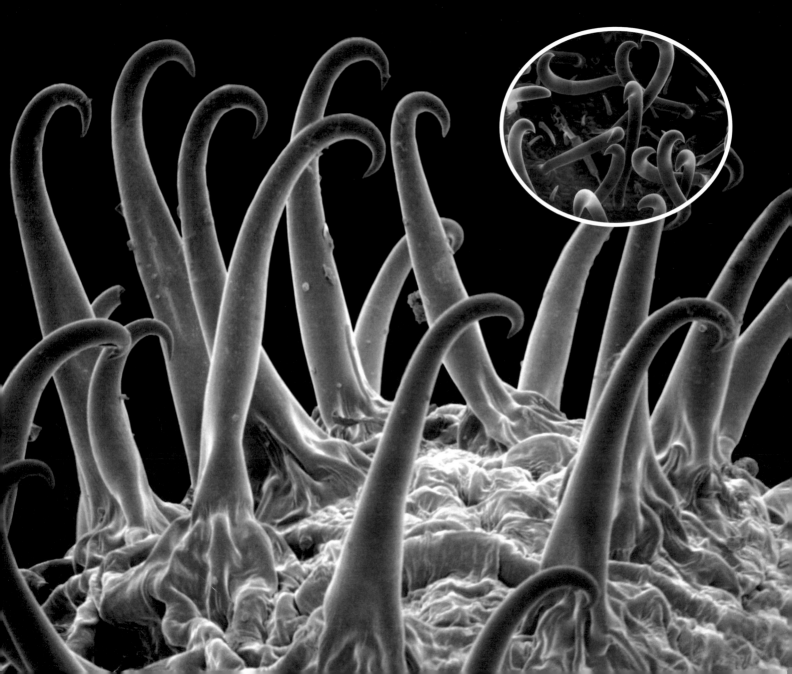

Burdock is a large green plant with broad leaves and flowers of pink or purple. It grows 1 to 5 feet (1 to 3m) tall with leaves that can be 18 inches (45cm) long. Its half-inch (1.27cm) flowers bloom during the summer. Deep in the centers of the flowers are the eggs that will grow into seeds. At the end of summer the flowers' petals dry up and die, leaving behind hard-shelled seedpods full of ripe seeds. The seeds are **dormant** throughout the cold winter. Then, in the spring they sprout into new burdock plants.

Purple and pink burdock flowers bloom during the summer, leaving behind burs that grow into new plants in the spring.

Burdock Burs

Burdock seedpods, called burs, are different from other kinds of seedpods. They are covered with sharp little spines. The spiny, hard-shelled burs protect the delicate seeds as they mature.

Once the seeds are fully ripe, the burdock plant has to have a way to send its seedpods to a new place where the seeds can grow. Without some **dispersal** method, burdock burs would just drop to the ground when they are ripe, and the seeds inside would grow right underneath the parent plant. One patch of ground

Prickly Seedpod

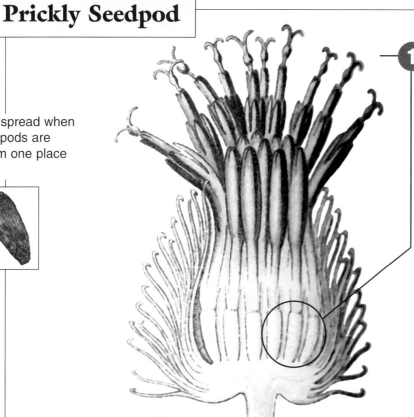

1 Burdock flowers bloom during summer. Young seeds begin to grow deep inside the pod.

2 After blooming, the seedpod becomes a dry, prickly bur. The bur's tough outer shell protects the mature seeds inside.

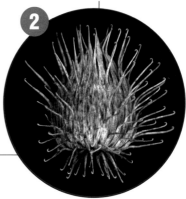

3 Hooked spines attach themselves to animals and people.

4 Seeds are spread when the prickly pods are carried from one place to another.

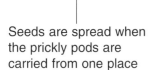

Burdock seeds hitch a ride on a bear's fur.
The inset shows a close-up of cockleburs
caught in deer fur.

would get very crowded with burdocks, and burdocks would not spread around very much. The bur is nature's clever solution to this problem.

Grabbing a Ride

Every burdock bur is a hitchhiker. It hitches a ride on living things that can walk. If an animal walks past a burdock plant and brushes against it, the burs fall off the plant and onto the animal's fur. The prickly, sticky burs latch on to the animal's fur and cling. Wherever the animal goes, the burs go along, too. Since one burdock can produce 16,000 seeds, burdock seeds get carried to many different places. When the animal brushes against other plants or trees, some burdock burs fall off and drop to the ground. If the landing place has good soil and the right amount of sunlight and water, a new burdock plant starts to grow.

Sticky and Smart

Burs are nature's way of dispersing burdock seeds and keeping the species thriving all over the world. Burdock seeds become common weeds because the burs can stick so well to their free rides.

Another Sticky Bur

Cockleburs, like burdocks, are members of the sunflower family and use burs for seed dispersal. The burs on a cocklebur are not round, but oval, like little footballs. Each bur has two seedsinside it. The burs are about 1 inch (2.5cm) long and are covered with barbed spines. Cocklebur burs are so sticky that some people use them to play a game of darts. Players throw the burs toward a dartboard made of cloth to see who can get the most bull's-eyes. The burs cling just as well as they do to animal fur or people's clothes.

Unlike round burdock burs, cockleburs (shown) are oval shaped.

Most people find burdock seeds a very irritating part of nature. As George de Mestral realized, burs are no fun to pick off dogs or clothes. But unlike other people, de Mestral was not just annoyed, he was curious. He wondered why burs were so successful as hitchhikers and what made them stick so well. He also wondered if he could imitate nature's smart invention.

Although most people think the burdock seed is a nuisance, de Mestral was fascinated with the sticky bur.

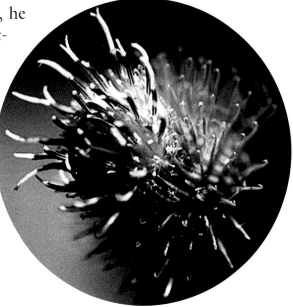

CHAPTER 2

Velcro!

George de Mestral was not only an outdoorsman. He was an inventor and engineer who designed useful things. When he was only twelve years old he designed a toy airplane and got a **patent** for it from the Swiss government. (The patent was proof that he had designed the plane. It meant that no one could steal his idea.) As an adult de Mestral had invented a new kind of asparagus peeler and a scientific device that measured the humidity in the air. On that summer day in 1948 he looked at the burs stuck in his dog's fur and started to think about a new invention. He grabbed a few burs and carried them inside his house to look at under a microscope.

Barbed Hooks

Peering into the microscope, de Mestral saw how the burs worked. The spines on each bur had tiny barbs or hooks on the tips that were the perfect size,

In the same way burdock hooks cling to fur, Velcro hooks (magnified, below right) cling to loops of fabric (above right).

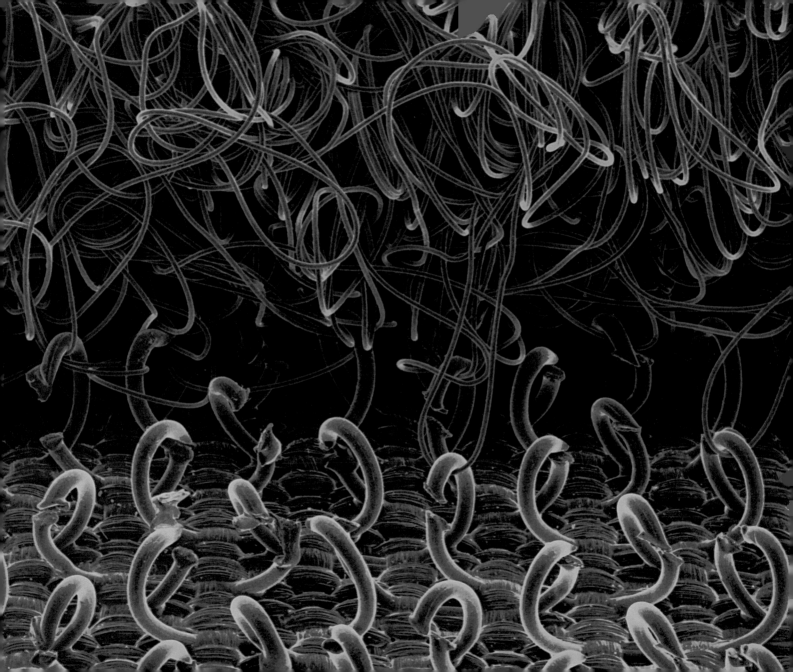

George de Mestral

George de Mestral was born in the Swiss village of Lausanne in 1907. He worked his way through college and became an electrical engineer. Then he served in his country's military and was a captain in the artillery during World War II. He was working in a machine shop when he first got his idea about hook and loop tape. Once his business became successful, de Mestral was a millionaire and used some of his wealth to help other young inventors to research and patent their ideas. He died in 1990. In 1999 he was inducted into the National Inventors Hall of Fame in honor of his valuable invention.

shape, and stiffness to grab onto the soft loops in clothes and animal fur. De Mestral was fascinated by this hook and loop system. He thought it was very practical. Perhaps he could design such a hook and loop system that would be a valuable and convenient fastener for people. The fastener would be stronger and easier to use than snaps, buttons, laces, or zippers. Later de Mestral talked to some friends about his idea, but almost no one took him seriously. Still, he was intrigued by the burs' sticking ability and determined to prove his idea could work.

Imitations in Cloth

De Mestral traveled to France in search of an expert weaver who could help him develop the new fastener. It would be made of two strips of cotton. One side would be the soft loop side, like animal fur, and the other would be the hook side, like burs. The loop side was easy, but the stiff hook side was hard to make. It took three years of experimenting to make a cotton fabric stiff enough that its loops could

From Barbs on a Weed to Velcro

be cut into hooks that would hold their shape. Then de Mestral and the weaver had to match the two sides. Sometimes they made the loops too big for the hooks. Other times the hooks were too big to latch on to the loops.

Finally, in 1951 de Mestral had two strips of fastener that stuck to each other when pressed together. He got a patent from the Swiss government and started a company to manufacture his invention. He called his fastener "hook and loop tape," and he named his company "Velcro." The name was a combination of the beginnings of two French words —*velour*, which means "velvet," and *crochet*, which means "hook."

De Mestral continued experimenting with hook and loop tape because the cotton strips still were not as sticky as burs. The strips pulled apart

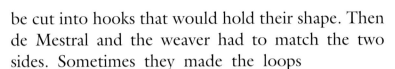

To create his hook and loop fastener, de Mestral experimented with cotton fabric (shown), which has looped fibers (insets).

VELCRO STRIP

A 5-inch (12.7cm) Velcro strip is so strong that it will not rip apart when holding a person who weighs 150 pounds (68kg).

Yet a 5-inch (12.7cm) Velcro strip is easily pulled apart with just two fingers when it is opened from one corner.

too easily. Then he discovered that nylon made better, stickier hook and loop tape. Nylon is not natural cloth, but plastic. When they were sheared under **infrared** heat, the microscopic loops melted a bit and became very strong hooks. In 1955 de Mestral got another patent for the nylon hook and loop tape and began selling his product.

Hook and Loop Tape

This hook and loop tape was so handy and practical that the Velcro company eventually became a huge success. The company set up plants in many different countries to produce hook and loop tape. The fastener was extremely popular. People thought of thousands of different uses for hook and loop tape, which they persisted in calling "Velcro" after the company name. De Mestral's curious idea turned into the world's favorite way to stick things together. Nature had indeed invented a perfect fastening system.

Velcro fasteners continue to be very popular with people all over the world.

CHAPTER 3

Thousands of Uses

Today Velcro Industries and many other companies make hook and loop fasteners. The original fastening system has been improved and changed so it can be used in different ways. Regular hook and loop fasteners keep sneakers on feet, fit baseball caps on different size heads, and snugly close cuffs on winter coats. Backpacks, watchbands, belts, ski boots, and inline skates seal tightly and securely with hook and loop tape. Hook and loop fasteners, however, are used for more than clothing in the modern world.

Clinging Anywhere

Researchers have developed fasteners much stronger than the nylon ones that de Mestral invented. Some fasteners are made with T-shaped hooks that have superior sticking power. Others are molded in plastic, stainless steel, and silver instead of cloth. These very tough fasteners have special purposes. Silver

A Creative Use

Kelly Werts is a musician who invented what he called Velcro tap dancing. He glued Velcro strips to the bottom of his shoes and danced on a carpet. The sound of his shoes being pulled from the carpet, like the sound of taps on hardwood, blended with the music in his shows.

Opposite: Velcro fasteners on both ends of this surfer's leash keep his board securely attached to his ankle.

21

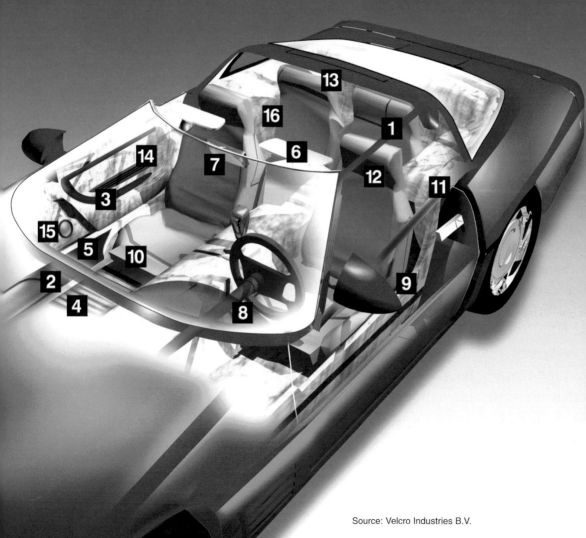

A Handy Invention: Velcro in a Sports Car

1 Headliners
2 Kick Panels
3 Armrests
4 Floor Mats
5 Map Pockets
6 Visor Mirrors
7 Overhead Consoles
8 Instrument Panels
9 Carpet
10 Seat Covers
11 Seat Belt Retainers
12 Headrests
13 Dome Lamps
14 Door Panels
15 Speaker Covers
16 Rear Seat Attachments

Source: Velcro Industries B.V.

hook and loop tape, for instance, is used with electrical wires. Silver fasteners do not interfere with the electricity being sent through the wires. T-shaped and steel fasteners attach liners to truck beds and are used in automobiles to hold body parts and seat covering in place.

Under Any Conditions

Researchers have also invented hook and loop fasteners that are waterproof, fireproof, and resistant to extreme weather conditions. Undersea divers use the waterproof fasteners as closures on their wetsuits and other gear. Aircraft fittings are made with the fireproof fasteners, as is firefighting equipment.

Soldiers wear uniforms and carry equipment with specially made hook and loop fasteners. These fasteners work equally well in

Velcro fasteners on this blood pressure checker keep the device tightly fastened to the patient's arm during use.

hot desert temperatures and under freezing arctic conditions. Military hook and loop fasteners work between temperatures of –58°F (–50°C) and 284°F (140°C). Scientists have even invented a fastener that makes no noise when it is pulled apart, so that soldiers will be safe when they are in enemy territory.

For Any Problem

Scientists have not only made extra tough hook and loop fasteners, they have also invented one-piece plastic fasteners for medical use. Instead of one side of hooks and the other side of loops, these fasteners alternate rows of hooks and loops on the same piece of tape. One strip of this fastener can attach to itself. People with sprained wrists or ankles, for example, wear braces that attach with this new kind of fas-

An Idea Catches On

When George de Mestral first began selling his hook and loop tape, few people wanted to buy it. Then, the space industry began using his tape for spacesuits, and the clothing industry realized how practical it was. By the 1960s his Velcro company was making and selling about 60 million yards (55 million meters) a year.

These astronauts use Velcro fasteners to keep their instruments from floating away in the weightlessness of space.

tener. The braces can be adjusted to fit tightly and hold injured parts still until they are healed. Plastic one-piece hook and loop tape is also useful with sports safety equipment, like helmets and padding.

Even in Space

Velcro fasteners are even in outer space where they help astronauts live and work on the International Space Station. In space everything is weightless, and everything floats. Nothing stays put unless it is attached to something. The surfaces on the inside of the space station have Velcro fastening strips where astronauts can put things that must not be lost. Tools and dishes are made to be attached to walls and shelves with hook and loop tape. Without Velcro fasteners, the inside of the space station would be a mess. Objects would float everywhere, and no one would be able to find the tool he or she wanted. Hook and loop tape keeps the International Space Station neat and organized.

Versatile Velcro

No one knows what uses hook and loop fasteners may be put to next, on Earth or in outer space.

Playing with Velcro Fasteners

People use Velcro fasteners for convenience and safety, but Velcro can also be used for silly games. People can rent an inflatable Velcro wall for a party. Everyone puts on Velcro jumpsuits and then leaps for the Velcro wall, where they stick like flies to the wall, even upside down. Or they can compete in a Velcro obstacle course where everybody gets stuck, and the first to get unstuck wins the race.

Pretending he is Spiderman, a boy dressed in a suit of Velcro sticks to a Velcro wall.

Some scientists are looking for ways to make edible fastener which will seal messy foods like burritos. Other scientists are experimenting with hook and loop fasteners to attach limbs to robots. Still others are considering a kind of giant hook and loop tape as a way to dock small satellites with large satellites or ships in space.

Nature invented a very good fastener when it used the hook and loop system for seed dispersal. Those little burs from a weed inspired an invention that continues to find creative uses all over the world.

Glossary

burs: Prickly coverings or protective shells surrounding the seeds of certain plants, such as burdock or cocklebur.

dispersal: The process by which a plant spreads seeds. Some seeds are dispersed by wind, some by being swallowed and deposited in droppings by birds or other animals, and some by burs that travel on animals.

dormant: In a condition of biological rest during which a living thing does not grow or function.

infrared: An invisible light or radiation that has a longer wavelength than visible light and produces heat.

patent: A grant by a government to an inventor. It gives the inventor the right to make and sell the invention, and prevents anyone else from doing so.

For Further Exploration

Books

Charlotte Foltz Jones, *Accidents May Happen: Fifty Inventions Discovered by Mistake*. New York: Delacorte, 1998. In this humorous book, readers can learn about many inventions that resulted from accidents or lucky ideas. The origins of loaves of bread, cornflakes, dynamite, yo-yos, and the telephone are described. The National Inventors Hall of Fame is also discussed.

Royston Roberts and Jeanie Roberts, *Lucky Science:* New York: John Wiley, 1995. This book not only tells stories of inventions and discoveries throughout history, it also describes experiments that kids can do themselves.

Don L. Wulffson, *The Kid Who Invented the Popsicle*. New York: Puffin, 1997. Almost everything people commonly use has a story behind it. Read short stories about many inventions and discoveries throughout history from animal crackers to Velcro to the zipper.

Web Sites

Halloween Costume Ideas for Kids (www.fabriclink.com/velcro/home.html.) Velcro USA Inc. provides this site with all sorts of ideas for making costumes for kids. Costume ideas and directions include a cape, an alarm clock, and an angel outfit.

United States Patent and Trademark Office Kids Pages (www.uspto.gov/go/kids/index.html). The U.S. Government has created this special page for kids who are interested in inventions and how to patent them. Readers learn about funny inventions that have been patented, play games, and explore the mystery pages.

Velcro.com: KidZone (www.velcro.com/kidzone.html). This is Velcro USA's official Web site. Read the story of George de Mestral's invention, as well as many fun facts about hook and loop tape.

Wayne's Word: Ultimate and Painful Hitchhikers (http://waynesword.palomar.edu/plmay98.htm). This site is written for older students, but it has many fascinating pictures of nature's stickiest and most painful burs. Some burs are so dangerous that they can cling to people's skin. Others are rated by "Sock Removal Difficulty Units" as the worst in the world to get off your clothes.

 Index

Picture Credits